Cambridge Elements ≡

Elements in Geochemical Tracers in Earth System Science
edited by
Timothy Lyons
University of California
Alexandra Turchyn
University of Cambridge
Chris Reinhard
Georgia Institute of Technology

IRON FORMATIONS AS PALAEOENVIRONMENTAL ARCHIVES

Kaarel Mänd
University of Alberta; University of Tartu

Leslie J. Robbins
University of Regina

Noah J. Planavsky
Yale University

Andrey Bekker
University of California, Riverside; University of Johannesburg

Kurt O. Konhauser
University of Alberta

CAMBRIDGE
UNIVERSITY PRESS

CAMBRIDGE
UNIVERSITY PRESS

University Printing House, Cambridge CB2 8BS, United Kingdom

One Liberty Plaza, 20th Floor, New York, NY 10006, USA

477 Williamstown Road, Port Melbourne, VIC 3207, Australia

314–321, 3rd Floor, Plot 3, Splendor Forum, Jasola District Centre,
New Delhi – 110025, India

103 Penang Road, #05–06/07, Visioncrest Commercial, Singapore 238467

Cambridge University Press is part of the University of Cambridge.

It furthers the University's mission by disseminating knowledge in the pursuit of
education, learning, and research at the highest international levels of excellence.

www.cambridge.org
Information on this title: www.cambridge.org/9781108995290
DOI: 10.1017/9781108993791

© Kaarel Mänd, Leslie J. Robbins, Noah J. Planavsky, Andrey Bekker, and Kurt
O. Konhauser 2021

First published 2021

A catalogue record for this publication is available from the British Library.

ISBN 978-1-108-99529-0 Paperback
ISSN 2515-7027 (online)
ISSN 2515-6454 (print)

Additional resources for this publication at www.cambridge.org/ironformations

Iron Formations as Palaeoenvironmental Archives

Elements in Geochemical Tracers in Earth System Science

DOI: 10.1017/9781108993791
First published online: December 2021

Kaarel Mänd
University of Alberta; University of Tartu

Leslie J. Robbins
University of Regina

Noah J. Planavsky
Yale University

Andrey Bekker
University of California, Riverside; University of Johannesburg

Kurt O. Konhauser
University of Alberta

Author for correspondence: Kaarel Mänd, kaarel.mand@ut.ee

Abstract: Ancient iron formations – iron- and silica-rich chemical sedimentary rocks that formed throughout the Precambrian aeons – provide a significant part of the evidence for the modern scientific understanding of palaeoenvironmental conditions in Archaean (4.0–2.5 billion years ago) and Proterozoic (2.5–0.539 billion years ago) times. Despite controversies regarding their formation mechanisms, iron formations are a testament to the influence of the Precambrian biosphere on early ocean chemistry. As many iron formations are pure chemical sediments that reflect the composition of the waters from which they precipitated, they can also serve as nuanced geochemical archives for the study of ancient marine temperatures, redox states, and elemental cycling if proper care is taken to understand their sedimentological context.

Keywords: iron formations, palaeotemperature, redox, Archaean, Palaeoproterozoic

ISBNs: 9781108995290 (PB), 9781108993791 (OC)
ISSNs: 2515-7027 (online), 2515–6454 (print)

Contents

1 Introduction

Iron formations (IF) are iron-rich (15–40 wt.% Fe) and siliceous (40–60 wt.% SiO_2) chemical sedimentary rocks (Figure 1) that precipitated from hydrothermally influenced seawater throughout the Precambrian aeons but with the majority of the preserved IF having been deposited between 2.80 and 1.85 billion years ago (Ga) in the Neoarchaean and Palaeoproterozoic eras (see Bekker et al. 2014; Konhauser et al. 2017 for reviews). IF formed in an array of environmental settings and, hence, display diverse textures and mineral compositions (Bekker & Kovalick 2021). Based on the geometry and tectonic setting of individual IF deposits, they are broadly divided into Algoma type and Superior type (Gross 1980), which in reality form the endmembers of a continuum (Bekker et al. 2012; Konhauser et al. 2017). The former are commonly described as smaller in extent, deposited in deeper-water conditions, and associated with local hydrothermal or volcanic activity. This hydrothermal or volcanic influence would have provided the dissolved Fe necessary for their formation. On the other hand, Superior-type IF formed more distally from Fe sources, are associated with diverse continental shelf sediments, and are more voluminous. In some instances Superior-type IF can be traced over distances of thousands of kilometres – they highlight the influence of reduced Fe(II)-bearing hydrothermal waters on Archaean and Palaeoproterozoic ocean chemistry (Konhauser et al. 2017).

Texturally, IF can be divided into those composed predominantly of granules (granular IF) and those consisting predominantly of fine-grained chemically precipitated muds (micritic IF) (e.g., Beukes & Gutzmer 2008). Micritic IF are composed of iron-rich muds and microcrystalline quartz (i.e., chert, composed of SiO_2). They often occur as distinctive, repetitive iron and chert layers, commonly termed 'banded IF'. The layering may be of variable thickness, from macrobands (metre thick) to mesobands (centimetre thick) to microbands (millimetre and submillimetre layers), the latter believed to represent an annual deposition process (Trendall & Blockey 1970; for a different view, see Bekker et al. 2012). Granular IF, instead, are characterised by granules of various sizes, composed of microcrystalline quartz, iron oxides, iron carbonates, and/or iron silicates that are cemented by chert, carbonate, or hematite. Most commonly, the granules were derived from local sedimentary reworking of earlier IF (Beukes & Gutzmer 2008). These different textures are indicative of specific sedimentary environments in which the IF formed. For instance, Beukes and Gutzmer (2008) divided the voluminous Superior-type Neoarchaean to Palaeoproterozoic IF in the Transvaal Supergroup, South Africa, into three main lithological facies. (1) Microlaminated chert-banded IF commonly formed in neutral-pH deepwater environments (hundreds of metres

Figure 1 (a) The 2.46-billion-year-old Brockman iron formation, Western Australia (Southern Ridge at Mount Tom Price mine). This is considered to rank among the largest known banded iron formations (IF) in the world. (b) Characteristic banding of the Dales Gorge Member of the Brockman iron formation. Photographs courtesy of Mark Barley. (c) Relative tonnage of IF in the preserved Precambrian rock record. Labels point out major contributions from individual iron formation deposits. Data from Bekker et al. (2014)

of water depth) where wave action had no effect; similar facies in Western Australia have been interpreted as having formed through submarine gravity flows carrying IF muds deposited in shallower settings (Krapež et al. 2003). (2) Chert-poor microlaminated IF with iron-silicate layers in South Africa were associated with somewhat shallower settings as they contain a component of fine-grained, storm wave–transported sediments. These must also have been precipitated in more alkaline, anaerobic water masses where iron-silicates were stable. (3) Micritic IF and granular IF lacking clear banding and sometimes rich in siderite were associated with shallower settings dominated by wave activity (Beukes & Gutzmer 2008).

Aside from textural classifications, IF may also be divided into sedimentary facies based on their mineralogy. In addition to chert, the best-preserved IF successions are composed of Fe oxides (e.g., magnetite (Fe_3O_4) and hematite (Fe_2O_3)), iron-rich silicate minerals (e.g., greenalite ($Fe_3Si_2O_5(OH)_4$)), or carbonate minerals (e.g., siderite ($FeCO_3$)), with locally sparse sulphides (e.g., pyrite (FeS_2)). The first three of these define the oxide, silicate, and carbonate facies of IF, respectively (James 1954). A sulphide facies postulated by James (1954) is now recognised to be more reflective of a shale facies and is not considered to be genetically related to IF (Bekker et al. 2010). It is generally agreed that none of the minerals in IF are primary in origin, in the sense that the original seafloor precipitate mineralogy was not preserved. Instead, the observed minerals reflect multiple post-depositional alteration events that occurred under both diagenetic (at low temperatures during progressive burial in sedimentary basins) and metamorphic (high-temperature transformations induced by tectonic and magmatic events) conditions. Nevertheless, it has been shown that preserved mineral assemblages change across a transect from deeper-water settings close to hydrothermal Fe(II) sources to shallower environments further away. For example, in the Mesoarchaean Witwatersrand IF in South Africa, this is expressed as a transition from hematite-dominated facies furthest from the shore to magnetite, and finally to siderite closest to the shore, reflecting increasing input of organic matter that enabled diagenetic Fe(III) reduction (Figure 2) (Smith et al. 2013). The specific form that this mineralogical succession takes depends on both global and local seawater, as well as sediment chemistry, and is in turn affected by basin geometry and biological factors (e.g., compare with Raye et al. 2015). It should also be pointed out that not all IF follow this mineralogical succession from deep to shallow. For instance, Wang et al. (2015) showed that in the 2.38–2.22 Ga Yuanjiacun IF,

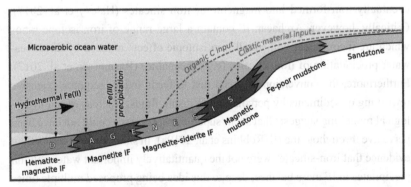

Figure 2 Mineralogical facies succession in the Mesoarchaean micritic Witwatersrand IF in South Africa (based on Smith et al. 2013) The IF are interpreted to have formed in relatively deepwater settings under a plume of hydrothermal Fe(II)-rich water, which was oxidised by chemolithoautotrophic bacteria to produce Fe(III) precipitates. During sediment diagenesis, Fe(III) was reduced to Fe(II), forming either magnetite or siderite, depending on the amount of available organic carbon as electron donor.

located in the North China Craton, in deeper waters more proximal to the hydrothermal vents, nutrients were abundant, and high biomass productivity was coupled to increased carbon burial, leading to the deposition of iron-rich carbonates. By contrast, hematite was deposited in shallow, well-oxygenated waters.

Traditionally, iron oxides in IF were interpreted to have formed from an initial ferric oxyhydroxide phase (e.g., ferrihydrite ($Fe(OH)_3$)), which precipitated in the photic zone from seawater via the oxidation of dissolved ferrous iron, Fe(II), at concentrations that may have ranged from 0.03 to 0.5 mM (e.g., Holland 1973). More recent work suggests that, in the presence of dissolved silica at concentrations estimated to have been as high as ~1 to 2 mM in the Archaean, the initial water column precipitate may instead have been a ferric oxyhydroxide-Si gel (e.g., Percak-Dennett et al. 2011). During IF mineralisation, the sinking of ferric oxyhydroxide-silica particles to the seafloor was followed by the formation of (1) magnetite or iron carbonates when the remineralisation of buried organic matter was coupled to Fe(III) reduction, either during diagenesis or metamorphism; (2) hematite, through dewatering and silica release, when organic carbon was insufficient for Fe(III) reduction; (3) iron silicates, such as greenalite, when silica sorbed onto or incorporated in ferric oxyhydroxides reacted with other cationic species within sediment pore waters (e.g., Morris 1993; Fischer & Knoll 2009); and (4) quartz, through dehydration of opaline silica and/or Fe(III)-Si gels. Ferrous iron sorption to these particles may also have given rise to 'green rust'-type deposits that eventually transformed into magnetite or iron silicates (Halevy et al. 2017). Critically, however, Archaean IF contain a large range of iron isotope ratios which are inconsistent with small iron isotopic effects induced by direct seawater precipitation of iron silicates (e.g., greenalite) (Rasmussen et al. 2017). Furthermore, the conversion of green rust to ferric oxides requires extensive reworking of sediments by percolating oxidising fluids, whereas recent hydrological modelling suggests that such 'supergene' processes could not have been pervasive throughout the IF (Robbins et al. 2019a). In short, there is compelling evidence that iron-silicates were not the quantitatively important water-column precipitates in IF mass balance, despite this idea being proposed multiple times recently (see Johnson et al. 2020 and references therein).

The generation of ferric iron minerals in IF depositional settings is generally ascribed to the metabolic activity of planktonic bacteria in ancient oceanic photic zones. The classic model invokes a redox-stratified ocean, with ferric iron precipitation occurring at the interface between oxygenated shallow waters and reduced upwelling ferrous iron-rich waters, the oxygen being sourced from cyanobacteria or their predecessors (Cloud 1973). These photoautotrophs would have flourished when nutrients were available and passively induced the precipitation of Fe(III) oxyhydroxide through their metabolic activity. Alternatively, anoxygenic photoautotrophic Fe(II)-oxidising bacteria (known as photoferrotrophs) could have directly oxidised Fe(II) and coupled this to the fixation of carbon-utilising light energy without any need for O_2 availability (Hartman 1984); this process may have accounted for most, if not all, Fe(III) deposited in IF (Konhauser et al. 2002). Furthermore, Kappler et al. (2005) calculated that the photoferrotrophs could have oxidised all hydrothermally derived Fe(II) before it reached surface waters, even if cyanobacteria were present in the oxic layer above – though iron isotope ratios of IF suggest that partial iron oxidation prevailed in Archaean and, to a lesser degree, Palaeoproterozoic oceans (Konhauser et al. 2017). Many aspects of the biology of cyanobacteria and photoferrotrophs seem to agree with an enhanced role of the latter – photoferrotrophs would have had a competitive advantage over early cyanobacteria since the former were better poised to benefit from upwelling phosphorous-rich deepwaters, given their adaptation to low-light conditions, especially since cyanobacteria have higher phosphate requirements (Jones et al. 2015). Furthermore, the ferruginous conditions associated with IF deposition may have been toxic to cyanobacteria (Swanner et al. 2015). Quantifying the relative contributions and extent of photoferrotrophs and cyanobacteria to IF deposition remains an area of ongoing research (Konhauser et al. 2018).

It should, however, be acknowledged that the story on IF deposition can be more complicated. For instance, IF deposition has also been attributed to mixing of discrete hydrothermally derived Fe(II) and Si-rich water plumes with sea-water possibly containing photosynthetic oxygen diffusing or downwelling from the photic zone. In cases where Fe(III) or Mn phases are preserved in sediments, deposition may also have occurred on the lower margins of the hydrothermal plume in contact with microaerobic water in a complex mixing configuration (Smith et al. 2013; Beukes et al. 2016). Those authors favoured this model because of the difficulty in explaining the retention of these oxidised phases as they settled through Fe(II)-rich bottom seawater. If true, this means that the depth and nature of Fe(III) precipitation, as well as the relative influence of hydrothermal water versus ambient seawater, were likely to have been highly variable in space and time, depending on the strength of both of these fluxes (Beukes & Gutzmer 2008). Moreover, Thompson et al. (2019) recently demon-strated with certain photoferrotrophic cultures that the ferric iron precipitates (ferrihydrite) did not attach to the cell surfaces such that ferrihydrite and biomass were not deposited together. This is important because this could lead to large-scale sedimentation of IF lean in organic matter, with excess biomass being deposited in coastal sediments elsewhere (i.e., forming organic-rich shales).

2 Iron Formations as Geochemical Proxies

IF are ubiquitous throughout the Archaean record and through the early part of the Palaeoproterozoic, beginning with the oldest known (metamorphosed) sedimentary succession in the world, the >3.77 Ga Nuvvuagittuq Supracrustal Belt, Canada (Mloszewska et al. 2012) (Figure 1c). Micritic IF remained abundant until around 1.85 Ga after which they declined in number due to a variety of environmental, tectonic, and magmatic factors. In contrast, granular IF first appeared in the rock record at ca. 2.90 Ga (Smith et al. 2017), reached their acme at ca. 1.88 Ga, and were then succeeded by smaller-scale IF, often lacking a chert component, in the Neoproterozoic and Phanerozoic record (after ca. 700 Ma) (Bekker et al. 2014). The other peak in the deposition of IF occurred in association with the Neoproterozoic glaciations – the so-called Snowball Earth events – when a dissolved iron reservoir was built in the oceans under ice cover either via hydrothermal supply or anoxic dissolution of reactive iron from sediments (see Bekker et al. 2014).

With a 2-billion-year record, IF have helped constrain both the redox conditions and transitions in the atmosphere-ocean system, as well as the composition of palaeo-seawater and its relationship to the evolution of the marine biosphere (see Robbins et al. 2016; Konhauser et al. 2017 for reviews). The utility of IF is based on a number of assumptions and conditions. First, ferrihydrite, the likely precursor phase for hematite and magnetite in IF, can faithfully preserve in the mineral the rare earth element (REE) distribution patterns of the mixture of seawater and hydrothermal water from which the IF precipitated (Bau & Dulski 1996) and, more broadly, other elemental and isotopic compositions of water, as has been demonstrated through ferrihydrite adsorption and diagenesis experiments (e.g., Døssing et al. 2011; Robbins et al. 2015; however, see also Halevy et al. 2017 for a different view). Second, this signal is commonly uncontaminated by continentally derived detrital materials, given that IF contain very low levels of detrital tracer elements such as aluminium and titanium. This means that IF geochemical data often provide a purely 'authigenic' record of marine chemistry. Finally, the generally low permeability and element mobility in IF mean that these geochemical signals have the potential to survive even high levels of metamorphism and are unlikely to have experienced widespread alteration by secondary fluids, the exception being later-stage ore formation (Frost et al. 2006; Robbins et al. 2015).

Yet a number of pitfalls complicate the straightforward interpretation of IF proxies, the foremost among them being the recent suggestion that the ferric iron-containing minerals now preserved in IF formed through post-depositional oxidation of primary Fe(II)-silicate minerals (e.g., Rasmussen et al. 2017; Muhling & Rasmussen 2020). In these models, the geochemical signals stored within IF would closely track basinal oxidising fluids rather than seawater (see aforementioned objections). Second, the straightforward back-calculation of isotope ratios and element concentrations in seawater based on IF data is often overly simplistic – for example, elemental adsorption coefficients and isotope fractionations involved in the precipitation of ferrihydrite are often known only through empirical observations (Konhauser et al. 2007) and may not fully capture the complex interplay among different components in solution, such as competing ions and dissolved organic compounds (see Robbins et al. 2016). This difficulty is compounded by a lack of direct modern analogues to IF that would allow us to study such mechanisms in the natural environment – a challenge not faced by the carbonate or shale records. Hydrothermal exhalative deposits, ironstones, and marine iron-manganese-oxide crusts may help to bridge this gap (e.g., Goto et al. 2020) even though these lithologies are genetically different from IF. Third, inefficient adsorption of some elements leading to low authigenic concentrations means that even minor addition of volcanic ash or detritus may overprint seawater chemical signals in IF (Haugaard et al. 2016; Thibon et al. 2019). In these cases, it is imperative that the mineral host of the studied chemical component be established and mixing relationships with detritus investigated. Finally, caution must be taken before interpreting IF chemical signals as that of open seawater since IF deposition may have taken place not in a homogenously stratified ocean but in a mixture of hydrothermally derived plume water and seawater, as evidenced by the mineralogical complexity of preserved IF (e.g., Smith et al. 2013) and widespread Eu anomalies indicating hydrothermal influence (e.g., Planavsky et al. 2010a). Even more, the deposition of many IF – especially, but not exclusively, Algoma type – was likely to have occurred in partially enclosed basins where water chemistry may have differed from that of the open ocean (Hoffman 1987; Beukes & Gutzmer 2008; Bekker et al. 2010). Hence it becomes very important for the interpretation of geochemical data to first gain an understanding of basin-wide facies architecture and to develop depositional models.

Despite these complexities, both the concentrations and isotope ratios of numerous elements in IF have been utilised to track a variety of palaeoenvironmental characteristics including first-order trends in trace-element availability through time (Robbins et al. 2016). In the subsequent sections we provide examples of the type of palaeoenvironmental data that IF have offered insights on.

3 Palaeotemperature

Minor but predictable differences in the partitioning of isotopes of the same element among different phases (i.e., isotope fractionation) in nature underlie the utility of stable isotope geochemistry. Since isotope fractionations are dependent on temperature, knowing the difference between an element's isotope ratios in an aqueous (e.g., seawater) and solid (e.g., chert) phase allows for the back-calculation of temperature coincident to formation. However, other factors affecting isotopic composition may overwhelm this signal, not least among them the possibility that IF deposited from heterogeneous mixtures of hydrothermal fluids and seawater. For example, an increasing trend in the ratio of the two main silicon isotopes ($^{30}Si/^{28}Si$, normalised to a reference standard and reported as $\delta^{30}Si$) of Archaean and Proterozoic IF was initially attributed to decreasing palaeotemperature (Robert & Chaussidon 2006) but was later reinterpreted to reflect a changing mixture of different Si sources and sinks bearing different $\delta^{30}Si$ values (Trower & Fischer 2019).

The most utilised palaeothermometer in chemical sediments is the oxygen isotope system ($^{18}O/^{16}O$ or $\delta^{18}O$) (Urey 1947). However, sedimentary $\delta^{18}O$ values are also dependent on the composition of seawater, leading to a long-standing debate on whether secular change in sedimentary $\delta^{18}O$ values is driven by variations in seawater composition or temperature (Knauth 2005; Bindeman et al. 2016). The study of $\delta^{18}O$ values in the chert layers of IF extended this debate into the Archaean, producing a trend of increasing $\delta^{18}O$ values from the Archaean to the Phanerozoic (Figure 3) (Perry 1967; Knauth 2005). Recent work has identified a similar $\delta^{18}O$ trend in ferric oxyhydroxides (Figure 3), where fractionation is known to be less dependent on temperature, thereby supporting secular change in seawater composition as the cause (Galili et al. 2019). However, measurements of the minor isotope ^{17}O have recently become more accessible (Bindeman et al. 2018; Bao 2019) and suggest that a more complex interplay among seawater composition, temperature, diagenesis, metamorphic exchange, and emergence of landmasses has shaped the preserved $\delta^{18}O$ trend (e.g., Liljestrand et al. 2020).

4 Nutrient Availability

Among the most straightforward palaeoenvironmental information stored in IF is a first-order reflection of the concentration of different elements in seawater. Simply by measuring elemental concentrations in IF formed at different times,

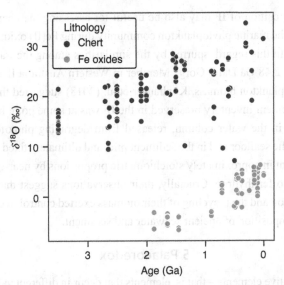

Figure 3 The oxygen isotope composition of chert and iron oxides through Earth history, which indicates either changing seawater composition or temperature. Chert (Robert & Chaussidon 2006) and iron oxide (Galili et al. 2019) data include iron formation samples; the offset in $\delta^{18}O$ values of these two phases is due to differing fractionation factors from seawater

temporal trends in the relative availability of nutrients or toxicity have been identified, although back-calculation to seawater concentrations can be complicated (see the previous section). For example, the IF record has been used to infer consistent and relatively high bottom water marine phosphorous (Planavsky et al. 2010b) and zinc (Robbins et al. 2013) concentrations in the Archaean and Palaeoproterozoic, but highly dynamic cobalt (Swanner et al. 2014) and nickel (Konhauser et al. 2009) trends with implications for ecosystems that utilise these nutrients. On the other hand, toxic elements like chromium (Konhauser et al. 2011) and arsenic (Chi Fru et al. 2019) show an increase in the early Palaeoproterozoic, suggesting that early organisms had to adapt to these new forms of toxicity.

A more nuanced tracer of nutrient cycling is the nitrogen isotope system ($\delta^{15}N$) – different N isotope ratios are indicative of different biological fixation and redox cycling pathways of this critical macronutrient (Stüeken et al. 2016). Nitrogen isotope data from highly metamorphosed IF of the earliest Archaean tentatively suggest the biological utilisation of hydrothermal nitrogen (Pinti et al. 2001), but Neoarchaean IF instead record the likely partial oxidation of ammonia to nitrate, indicating the onset of an oxic N cycle (Garvin et al. 2009; Busigny et al. 2013).

The composition of IF may also be useful for assessing the chemical composition of the marine phytoplankton community during Fe(II) oxidation and IF deposition. In this regard, spurred by the similarities among the trace-element ratios in the 2.48 Ga Dales Gorge Member in Western Australia IF and anoxygenic phytoplankton biomass, Konhauser et al. (2018) suggested that much of the trace-element inventory preserved in the IF was at some point biologically assimilated in the water column, released from degrading photoferrotrophic biomass at the seafloor and in the sediment pile, and ultimately fixed in the iron-rich sediment in approximately stoichiometric proportions by near-quantitative adsorption to ferrihydrite. Crucially, their observations suggest that, as today, phytoplankton and the recycling of their biomass exerted control over the trace-element composition of ancient seawater and sediment.

5 Palaeoredox

Redox-sensitive elements – that is, elements that occur in different valence states with contrasting chemical behaviour in natural environments – are a valuable tool for tracking the abundance of oxygen in Earth's atmosphere and oceans and, by extension, the emergence and prevalence of oxygenic photosynthesis (e.g., Farquhar et al. 2014). For instance, the IF record can be used to identify concentration trends of an element like uranium, which is significantly more soluble in oxygen-rich water, so that higher U concentrations in the IF record point to higher oxygen availability at that time in Earth's history (Partin et al. 2013). REEs constitute another widely applied redox proxy – this otherwise similarly acting set of lanthanide elements displays differing adsorption affinities to manganese oxides (e.g., Bau & Dulski 1996). Since manganese oxides only form in oxic conditions (see, e.g., Johnson et al. 2013; Daye et al. 2019 for alternate interpretations), specific anomalies in REE concentration patterns, namely negative cerium anomalies, attest to the presence of oxygen and redox-stratified water masses; the lack of such anomalies in Archaean IF signifies a predominantly anoxic world (Planavsky et al. 2010a).

The most recent palaeoredox work, instead, utilises redox-sensitive isotope systems. Perhaps the best example is the iron isotope system (δ^{56}Fe) (see Dauphas et al. 2017; Johnson et al. 2020). The isotope fractionation in ferric oxyhydroxides is always expected to be positive; that is, the precipitating phase is enriched in heavy Fe isotopes compared to seawater, regardless of the mechanism of Fe(II) oxidation to Fe(III) (see references in Dauphas et al. 2017). This fractionation, however, is only expressed if Fe(II) oxidation is partial, not quantitative (Rouxel et al. 2005). As such, predominantly positive δ^{56}Fe values in Archaean and Palaeoproterozoic IF indicate that ferric oxyhydroxides precipitated from water masses that remained Fe(II)-rich and oxygen-poor, so that only a portion of the dissolved Fe pool was precipitated as IF (Planavsky et al. 2012; Heard et al. 2020). Negative IF $\delta^{56}Fe$ values, as seen in many <3.0 Ga IF, may then reflect late-stage Fe precipitation from a reservoir that underwent extensive, prior drawdown of heavy Fe isotopes, resulting in isotope distillation (Rayleigh fractionation) in the dissolved Fe(II) pool (Planavsky et al. 2012). Alternatively, negative $\delta^{56}Fe$ values can reflect diagenetic overprinting of the primary IF sediment because the process of dissimilatory iron reduction (DIR) – a heterotrophic metabolism in which bacteria oxidise organic carbon and reduce ferric iron – leads to pore waters enriched in dissolved Fe(II) with negative $\delta^{56}Fe$ values that could reprecipitate as magnetite or siderite (Johnson et al. 2008; Heimann et al. 2010).

Molybdenum isotopes ($\delta^{98}Mo$) (Kendall et al. 2017) are most significantly fractionated by adsorption onto manganese oxides in oxic seawater (Barling & Anbar 2004). Therefore, the $\delta^{98}Mo$ composition of residual seawater scales with the global prevalence of oxic conditions in the oceans, allowing for quantification of the oceanic redox balance (Arnold et al. 2004). Compared to shales and carbonates, however, there are relatively large $\delta^{98}Mo$ fractionations onto ferric oxyhydroxides, making such quantitative assessments more difficult. Iron formation $\delta^{98}Mo$ values have instead been used for the less-precise task of identifying local Mn(IV) oxide formation from dissolved Mn(II), that is, the presence of oxygen oases in a predominantly anoxic world (e.g., Planavsky et al. 2014). The 'smoking gun' in these types of studies is an anti-correlation between the Fe/Mn ratio and $\delta^{98}Mo$ values (Figure 4) (Planavsky et al. 2014), which arises due to the different Mo isotope fractionation imparted by adsorption onto Fe(III) versus Mn(IV) oxyhydroxides (Wasylenki et al. 2008; Goldberg et al. 2009). Goto et al. (2020) recently proposed a quantitative model for estimating seawater $\delta^{98}Mo$ values from measured IF values combined with Fe/Mn ratios (Figure 4). IF may, therefore, ultimately prove to have a similar utility as shales and carbonates in assessing the global balance of seafloor redox conditions.

The 'stable' uranium isotope system ($\delta^{238}U$) is a more recently developed global palaeoredox proxy (Lau et al. 2019). The most significant $\delta^{238}U$ fractionation accompanies biotic reduction of soluble U(VI) to insoluble U(IV) in anoxic sediments, leaving the marine pool depleted in heavy U isotopes. Therefore, in parallel with $\delta^{98}Mo$, marine $\delta^{238}U$ negatively scales with the areal extent of anoxic conditions on the seafloor (Andersen et al. 2014). Uranium incorporation into IF is still poorly understood (Skomurski et al. 2011), but a comprehensive dataset of $\delta^{238}U$ values in IF was recently presented by Wang et al. (2018). They identified an increase in the variance of $\delta^{238}U$ ratios at 2.95 Ga – IF deposited following this date

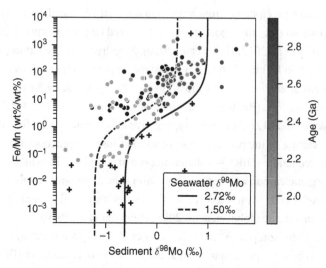

Figure 4 The effect of manganese versus iron oxide cycling on sediment molybdenum isotope composition, replotted from Goto et al. (2020). Circles represent Archaean and Palaeoproterozoic iron- or manganese-rich sediments, coloured per age; crosses represent modern marine Fe-Mn crusts. Lines represent two mixing models for Mo adsorbed to sedimentary iron and manganese oxides, starting from seawater δ^{98}Mo compositions of 2.72‰ (solid line, modern) and 1.50‰ (dashed line, Paleoproterozoic estimate)

had markedly negative δ^{238}U values relative to igneous background values, indicating active U(VI) redox cycling in the oceans.

The most consequential Cr isotope (δ^{53}Cr) fractionations occur on land during the oxidative weathering of Cr(III)-bearing minerals and the resultant production of aqueous Cr(VI) (Frei et al. 2009; Wei et al. 2020), though Mn(II)-oxidising bacteria can also mediate Cr(III) oxidation in aqueous environments (e.g., Murray et al. 2005). Chromium(III) oxidation and the transport of fractionated Cr(VI) to oceans require the generation of Mn(IV) oxides in weathering environments and the absence of Fe(II) in rivers, which translates to >0.01–0.2% O_2 in the atmosphere (e.g., Crowe et al. 2013). Fractionated δ^{53}Cr values in sediments could, therefore, be a proxy for atmospheric oxygenation (Frei et al. 2009). IF, which can capture seawater δ^{53}Cr values with fidelity (Døssing et al. 2011), have featured prominently in Cr isotope studies, wherein (generally positive) deviations from the igneous background δ^{53}Cr composition have been taken as a sign of transient oxygen abundance in the Archaean and Palaeoproterozoic atmosphere (Frei et al. 2009; Crowe et al. 2013). Critically, modern weathering in outcrops is capable of significantly affecting the primary Cr budget (Albut et al. 2018; Heard et al. 2021). In this light, multiple lines of evidence, based on stable and radiogenic isotope ratios as well as elemental abundances, should be considered in studies of early oxygenation rather than relying on single-proxy arguments.

6 Bulk versus In-situ Analysis

An important consideration when viewing the IF record is the value and underlying meaning of both bulk and in-situ analysis. Bulk analyses are often made by analysing trace-element concentrations or isotope ratios on a whole-rock dissolved sample whereas micro-scale in-situ analysis may be facilitated by either laser ablation or secondary ionisation mass spectrometry (SIMS). While both yield important information regarding the geochemical conditions contemporaneous to deposition and during diagenesis, the distinction is important.

With regards to trace-element concentrations, in many of the records produced thus far, including Ni (Konhauser et al. 2009), Cr (Konhauser et al. 2011), and Zn (Robbins et al. 2013), the highest concentrations recorded in IF are often captured by laser ablation analyses of the iron-bearing minerals. This broad observation may be the result of one of several underlying causes. First, whole-rock sampling of IF often includes the typically trace element-poor chert layers, which may lead to lower overall concentrations, while in-situ analyses capture the contrast in chert and Fe layers. Assuming that the iron- and silica-rich bands reflect distinct depositional conditions (e.g., Posth et al. 2008; Schad et al. 2019), whole-rock analyses may offer a time-integrated or averaged reflection of trace element availability during IF deposition. Second, differences in trace element concentrations observed between magnetite and hematite within a given IF may be the result of localised diagenetic effects, which could lead to preferential sequestration of a given trace metal within a specific iron phase. For instance, in a survey of several IF, Robbins et al. (2019b) found that the highest concentrations of Zn typically occurred in magnetite. Encouragingly, however, it was noted by the authors that mean values from laser-ablation analyses were typically within an order of magnitude of whole-rock data. This suggests an overall consistency with whole-rock analyses and demonstrates the inherent need for a combination of both whole-rock and in-situ analyses when considering the environmental implications of trace-element records.

Utilising in-situ laser ablation and SIMS analyses, multiple authors have reported a range of δ^{56}Fe (e.g., Steinhoefel et al. 2010; Li et al. 2013), δ^{18}O (e.g.,

Heck et al. 2011; Li et al. 2013), and δ^{30}Si (e.g., Steinhoefel et al. 2010; Heck et al. 2011) isotope ratios in IF, providing critical clues as to IF precipitation mechanisms and diagenetic processes. For instance, Steinhoefel et al. (2010) noted a relatively uniform Si isotope composition in cherts from the Hamersley Basin, Australia and Transvaal Supergroup, South Africa, suggesting precipitation from well-mixed seawater, while a range in δ^{56}Fe values from -2.2 to $0‰$ in magnetite was interpreted to reflect the formation of magnetite through post-depositional processes fuelled by DIR. Similarly, Li et al. (2013) described systematic differences in δ^{56}Fe values between hematite and magnetite grains and argued that while both were ultimately derived from a common precursor phase, the difference is attributable to reductive processes during early diagenesis. In the same contribution, Li et al. (2013) provided δ^{18}O evidence that hematite present in the Dales Gorge Member of the Brockman iron formation, Australia, is the result of interaction with metamorphic fluids and dehydration of $Fe(OH)_3$.

While in-situ studies have noted key differences in the Fe isotope composition of the iron-bearing minerals in IF, they have supported the overall inferences gained from whole-rock Fe isotope analyses, whereby the more positive δ^{56}Fe values represent the precipitation of a ferric precursor mineral via partial oxidation and the more negative phases such as magnetite formed through subsequent diagenetic, redox-driven modifications (see Dauphas et al. 2017).

7 Comparison of IF, Shale, and Carbonate Mo Records: A Case Study

The IF record has thus far provided significant insights, but moving forward, a potentially fruitful pathway to glean even more information pertaining to the early Precambrian marine environments is to compare and contrast elemental concentration and isotope data from various marine sediments, including black shales and carbonates in addition to IF. Black shales comprise fine-grained clastic sediments with >1 wt.% of carbonaceous material that formed under anoxic or euxinic (H_2S-rich) water-column conditions, and they represent a large fraction of all ancient marine sediments preserved today. Like IF, they often contain authigenic enrichment of trace elements whose sedimentary history testifies to major biogeochemical events related to the development of aerobic ecosystems and the rise of atmospheric oxygen (e.g., Scott et al. 2008). Unlike the deeper-water black shales, a number of Precambrian marine carbonates have formed on shallow-marine platforms, and as such their composition reflects the palaeoenvironmental conditions of shallow-water and detrital sediment-starved settings (e.g., Petrash et al. 2016). Although each lithology has been used with some success to decipher temporal variations in seawater composition, relatively little attention has been paid towards integration of long-term datasets of different lithologies to provide a comprehensive picture of the palaeonutrient landscape that, as today, should have varied along an ocean transect. To demonstrate its potential, we use a combination of Mo datasets compiled for each lithology to compare temporal trends through the transition from the low-oxygen Mesoarchaean world to the post-Great Oxidation Event (GOE) period (Figure 5).

The record of Mo concentrations through time in shales was initially assembled by Scott et al. (2008), with the aim of tracking the onset and protracted nature of Earth's long-term oxygenation. Shale records show the first enrichments in Mo beginning after about 2.8 Ga, but increasing more rapidly in the immediate lead up to the GOE at 2.5 Ga (Figure 5a). Despite being sparse, the carbonate record for Mo (Figure 5c) (Thoby et al. 2019) is, overall, consistent with such a trend: lower concentrations at about 3.0 Ga are followed by, on average, higher concentrations between 2.6 and 2.5 Ga. To date, however, a compilation of Mo concentrations through time in IF has yet to be examined, as most of the information gained from Mo in IF has been through isotopic studies (see Section 5). A compilation of such IF data (Figure 5 source data; Figure 5e) is characterised by low Mo abundances at ca. 2.6 Ga, and higher average values at ca. 2.5 to 2.3 Ga, coinciding with the early stage of the GOE. While the three records are consistent in terms of general trends, the onset of increase in Mo concentrations in IF appears to lag behind that in shales and carbonates. We propose that this may be due to the generally more distal (i.e., further from the shore) and deeper-water setting of IF deposits of the Archaean – Mo, solubilised from continents through incipient oxidative weathering in the lead up to the GOE, was quickly sequestered in relatively near-shore, anoxic environments, recorded by carbonates and shales. Molybdenum concentrations in distal IF depositional sites only began to increase once significant atmospheric oxygen levels increased Mo input into the oceans and overwhelmed the ability of ocean margins to effectively sequester all Mo delivered from the continents.

A comparative dataset of Mo isotope compositions across these carbonates and shales (first compiled by Thoby et al. 2019) shows, as expected, a trend mostly resembling those of Mo concentrations – relatively small fractionations in the Archaean lead up to a rise at 2.6 Ga preceding the GOE, indicating the onset of more extensive redox cycling of Fe and Mn (Figure 5b, d). Interestingly, fractionated Mo is also found in the Mesoarchaean carbonate record, suggesting redox cycling in the shallow carbonate-forming environments, whereas any such signal was masked by detrital input to the shales.

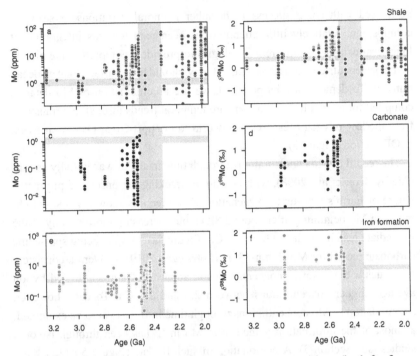

Figure 5 Concentrations (a, c, e) and isotope compositions (b, d, f) of molybdenum in shales (a, b), carbonates (c, d), and iron-rich rocks (e, f) from the Mesoarchaean to the Palaeoproterozoic. Grey vertical bars represent the Great Oxidation Event, as defined through sulphur isotope systematics (Gumsley et al. 2017; Warke et al. 2020). Grey horizontal bars represent the average crustal composition (Rudnick & Gao 2014; Breillat et al. 2016). Circles represent whole-rock or leach analyses; crosses on panel e are in-situ measurements. Concentration and isotope data for shales are from the compilations of Robbins et al. (2016) and Thoby et al. (2019), as well as from studies by Asael et al. (2018), Ossa Ossa et al. (2018), Cabral et al. (2019), Ostrander et al. (2019, 2020), and Mänd et al. (2020). Carbonate data are from the compilation of Thoby et al. (2019). Iron formation data are compiled in this study (Figure 5 source data), but also include data from Thoby et al. (2019), and Albut et al. (2019)

However, it is worth noting that Mo ores – the products of high-temperature processes – can also be marked by large variability in Mo isotope values (Mathur et al. 2010) and may, therefore, contribute fractionated detrital material to some sediments. The current IF δ^{98}Mo record shows less consistency and lacks a clearly increasing trend in the lead up to the GOE, which may be an artefact of the small sample size (Figure 5f). However, the presence of fractionated values throughout the Archaean agrees with the carbonate record and suggests subtle local redox cycling captured in the authigenic IF record. While limited in existing datasets, the post-GOE data across all lithologies indicate stability in the Mo cycle for the next few hundred million years.

Overall, the comparison of Mo isotope records across these three lithologies suggests a Mesoarchaean world with faint oxygen oases and Mn redox cycling both in shallow- and deeper-water environments, the signals of which are captured in the Mo isotope records of authigenic sediments like carbonates and IF, but which were globally overwhelmed by an anoxic, detrital Mo cycle reflected by the shale record. The situation gradually transformed in the Neoarchaean, as oxygen oases, or transient global oxygenation events, became more frequent, and an oxidative weathering flux of Mo from the continents reached the depositional settings of both shallow-water carbonates and shales, without affecting deep-ocean basins. Only during and after the GOE did pervasive oxygenation finally manage to extend the oxidative Mo cycle to the environments where IF were deposited.

8 Outlook

Examinations of the IF record have benefited greatly from advancements in analytical instrumentation, with non-traditional stable isotope analyses becoming increasingly prevalent in studies over the last several years. Stable isotope systems in IF that are beginning to see a great deal of interest include the trace metals Ni, Zn, Cu, Se, Ce, and Ge. In the case of nickel – a critical trace nutrient for methane-producing microbes – studies are being focused on understanding the fractionation of Ni isotopes during adsorption onto Fe-oxyhydroxides with an eye towards how that will affect the interpretation of the dynamic IF-hosted Ni record (Gueguen et al. 2018). A rise in zinc isotopic ratios over the GOE has controversially been proposed to reflect the onset of continental weathering of phosphate, which may have complexed with Zn in aqueous environments resulting in isotope fractionation (Pons et al. 2013). But more recent studies are, instead, interested in the strong isotope effects associated with biotic uptake of Zn, given the biological importance of this trace nutrient, particularly to eukaryotes (Moynier et al. 2017). Copper isotopic ratios in black shales display a shift from negative to positive values over the GOE, which has been explained by increasing oxidative supply of weathering-derived isotopically heavy Cu after 2.3 Ga along with the waning of extensive IF deposition, which preferentially removes heavy Cu isotopes (Chi Fru et al. 2016). However, this hypothesis has yet to be confirmed through IF studies, which are hampered by low authigenic concentrations of Cu, leading to contamination issues (Thibon et al.

2019). Future IF data will also help refine our understanding of the selenium (Xu et al. 2020) and cerium (Nakada et al. 2016) isotope systems, both of which may act as nuanced proxies for hypoxic aqueous environments, and the germanium isotope system, which has been proposed to track the extent of submarine hydrothermal processes and continental weathering (Rouxel & Luais 2017). Moving forward, these isotopic systems will undoubtedly yield powerful new insights into the coupled evolution of the Earth's ancient oceans and life that is recorded in the chemistry of IF.

Other major advances over the next few years will likely be focused on resolving the ongoing debate about primary mineralogy and the effects of diagenesis. Iron silicates have been increasingly invoked as the primary mineralogy based on detailed petrographic observations (Muhling & Rasmussen 2020) – despite that this is contradicted by IF carbon and iron isotope records (e.g., Craddock & Dauphas 2011). Resolving this debate is essential for an understanding of the basics of IF deposition and how we can use these unique rocks as palaeoenvironmental archives of past environments. The coupling of stable and radiogenic isotope toolkits (e.g., Albut et al. 2018) is one promising way forward to gauge the effect of secondary alteration. Radiogenic isotope studies can provide compelling evidence that the studied sedimentary rocks have been 'closed' to alteration since deposition with respect to the isotopic system, indicating their potential as a biogeochemical archive. Alternatively, evidence for an open-system behaviour can be a clear red flag that the sedimentary rocks are not ideally suited to understand the evolution of the Earth System (e.g., Hayashi et al. 2004). This is especially relevant as differences in geochemistry between outcrop and core IF samples from the same formations have been noted, and these differences provide contrasting interpretations of ancient redox signals (Albut et al. 2018; Heard et al. 2021). Outcrop samples are indeed more susceptible to modern weathering effects than previously thought. While no large-scale assessment of the IF record has been made in this light, a statistical investigation of cerium anomalies in shales through time has shown a clear bias between outcrop and core samples (Planavsky et al. 2020).

Finally, a key concern in the present IF research landscape is that advances in geochemical methods have far outpaced advances in field- and drill core-based sedimentology. As a result, geochemical results are often reported from haphazardly selected samples, with very little consideration of their particular sedimentological context, but which can significantly affect the geochemical interpretations. A way forward would require building integrated geochemical and sedimentological models, which acknowledge that ancient IF depositional settings were just as complex as modern sedimentary systems. Doing so will be difficult without increased effort is put into finding support for thorough scientific drilling programs that can answer questions about broader basin architecture, as well as redox and chemical stratification. Carefully selected drill-core material is also necessary to avoid major problems with recent meteoric alteration of geochemical signals that have recently been highlighted for IF outcrop samples (Albut et al. 2018).

Key Papers

1. James (1954) – classic petrographic description and classification of IF.
2. Perry (1967) – established the rising trend in $\delta^{18}O$ values through the Archaean and Proterozoic IF–chert record.
3. Cloud (1973) – a pioneering article on the biological significance in precipitation of IF.
4. Bau & Dulski (1996) – classic paper showing the preservation of seawater-like REE patterns and how these can be used to determine REE sources to IF.
5. Konhauser et al. (2002) – quantitatively shows that bacteria could have played a major role in the precipitation of IF.
6. Krapež et al. (2003) – a perspective on the facies of IF being deposited as resedimented density currents of iron-rich hydrothermal muds, iron oxy-hydroxides, and siderite.
7. Kappler et al. (2005) – demonstrated the ability of photoferrotrophs to precipitate IF.
8. Rouxel et al. (2005) – revealed the major trends in IF $\delta^{56}Fe$ values through Earth history. The first among a multitude of modern studies targeting redox proxies in IF time series.
9. Beukes & Gutzmer (2008) – an in-depth overview of the complex facies associations of the major Ghaap-Chuniespoort Group and Transvaal Supergroup IF of South Africa, including an up-to-date textural and facies classification of Superior-type IF.
10. Bekker et al. (2014) – key modern review paper on IF.
11. Muhling & Rasmussen (2020) – showed that greenalite is present in nearly every IF, lending support for the controversial hypothesis that IF were originally precipitated as Fe(II)-silicate phases and only later were oxidised to Fe(III).

Supplementary Data

Additional resources for this publication at www.cambridge.org/ironformations

1. Source data for Figure 1c. Age in Ga, tonnage in Gt.
2. Source data for Figure 3. Age in Ga; $\delta^{18}O$ in per mil (‰), normalised to Vienna Standard Mean Ocean Water. Errors are reported to different standards, as given in literature sources.
3. Source data for Figure 4. Age in Ga; $\delta^{98}Mo$ in per mil (‰), normalised to NIST SRM 3134 = 0.25‰; Fe/Mn in wt.% over wt.%. Errors are reported to different standards, as given in literature sources.
4. Source data for Figure 5. Age in Ga; $\delta^{98}Mo$ in per mil (‰), normalised to NIST SRM 3134 = 0.25‰; Mo concentrations in ppm. Errors are reported to different standards, as given in literature sources.

References

Albut, G., Babechuk, M. G., Kleinhanns, I. C., et al. (2018). Modern rather than Mesoarchaean oxidative weathering responsible for the heavy stable Cr isotopic signatures of the 2.95 Ga old Ijzermijn iron formation (South Africa). *Geochimica et Cosmochimica Acta*, **228**, 157–189.

Albut, G., Kamber, B. S., Brüske, A., et al. (2019). Modern weathering in outcrop samples versus ancient paleoredox information in drill core samples from a Mesoarchaean marine oxygen oasis in Pongola Supergroup, South Africa. *Geochimica et Cosmochimica Acta*, **265**, 330–353.

Andersen, M. B., Romaniello, S., Vance, D., et al. (2014). A modern framework for the interpretation of 238U/235U in studies of ancient ocean redox. *Earth and Planetary Science Letters*, **400**, 184–194.

Arnold, G. L., Anbar, A. D., Barling, J., & Lyons, T. W. (2004). Molybdenum isotope evidence for widespread anoxia in mid-Proterozoic oceans. *Science*, **304**(5667), 87–90.

Asael, D., Rouxel, O., Poulton, S. W., et al. (2018). Molybdenum record from black shales indicates oscillating atmospheric oxygen levels in the early Paleoproterozoic. *American Journal of Science*, **318**(3), 275–299.

Bao, H. (2019). Triple oxygen isotopes. In T. Lyons, A. Turchyn & C. Reinhard (Eds.), *Elements in Geochemical Tracers in Earth System Science*, Cambridge University Press, Cambridge, UK.

Barling, J., & Anbar, A. D. (2004). Molybdenum isotope fractionation during adsorption by manganese oxides. *Earth and Planetary Science Letters*, **217** (3–4), 315–329.

Bau, M., & Dulski, P. (1996). Distribution of yttrium and rare-earth elements in the Penge and Kuruman iron-formations, Transvaal Supergroup, South Africa. *Precambrian Research*, **79**(1), 37–55.

Bekker, A., & Kovalick, A. (2021). Ironstones and iron formations. In D. Alderton & S. A. Elias, eds., *Encyclopedia of Geology* (2nd ed.), Oxford: Academic Press, pp. 914–921.

Bekker, A., Krapež, B., Slack, J. F., et al. (2012). Iron formation: The sedimentary product of a complex interplay among mantle, tectonic, oceanic, and biospheric processes – a reply. *Economic Geology*, **107**, 379–380.

Bekker, A., Planavsky, N. J., Krapež, B., et al. (2014). Iron formations: Their origins and implications for ancient seawater chemistry. In H. D. Holland & K. K. Turekian, eds., *Treatise on Geochemistry* (2nd ed.), Oxford: Elsevier, pp. 561–628.

Bekker, A., Slack, J. F., Planavsky, N., et al. (2010). Iron formation: The sedimentary product of a complex interplay among mantle, tectonic, oceanic, and biospheric processes. *Economic Geology*, 105(3), 467–508.

Beukes, N. J., & Gutzmer, J. (2008). Origin and paleoenvironmental significance of major iron formations at the Archean-Paleoproterozoic boundary. In S. Hagemann, C. A. Rosière, J. Gutzmer, & N. J. Beukes, eds., *Banded Iron Formation-Related High-Grade Iron Ore*, Vol. 15, Society of Economic Geologists Littleton, CO, USA, pp. 5–47.

Beukes, N. J., Swindell, E. P. W., & Wabo, H. (2016). Manganese deposits of Africa. *Episodes Journal of International Geoscience*, 39(2), 285–317.

Bindeman, I. N., Bekker, A., & Zakharov, D. O. (2016). Oxygen isotope perspective on crustal evolution on early Earth: A record of Precambrian shales with emphasis on Paleoproterozoic glaciations and Great Oxygenation Event. *Earth and Planetary Science Letters*, 437, 101–113.

Bindeman, I. N., Zakharov, D. O., Palandri, J., et al. (2018). Rapid emergence of subaerial landmasses and onset of a modern hydrologic cycle 2.5 billion years ago. *Nature*, 557, 545–548.

Breillat, N., Guerrot, C., Marcoux, E., & Négrel, Ph. (2016). A new global database of δ^{98}Mo in molybdenites: A literature review and new data. *Journal of Geochemical Exploration*, 161, 1–15.

Busigny, V., Lebeau, O., Ader, M., et al. (2013). Nitrogen cycle in the Late Archean ferruginous ocean. *Chemical Geology*, 363, 115–130.

Cabral, A. R., Zeh, A., Vianna, N. C., et al. (2019). Molybdenum-isotope signals and cerium anomalies in Palaeoproterozoic manganese ore survive high-grade metamorphism. *Scientific Reports*, 9(1), 4570.

Chi Fru, E., Rodríguez, N. P., Partin, C. A., et al. (2016). Cu isotopes in marine black shales record the Great Oxidation Event. *Proceedings of the National Academy of Sciences of the United States of America*, 113(18), 4941–4946.

Chi Fru, E., Somogyi, A., Albani, A. E., et al. (2019). The rise of oxygen-driven arsenic cycling at ca. 2.48 Ga. *Geology*, 47(3), 243–246.

Cloud, P. (1973). Paleoecological significance of the banded iron-formation. *Economic Geology*, 68(7), 1135–1143.

Craddock, P. R., & Dauphas, N. (2011). Iron and carbon isotope evidence for microbial iron respiration throughout the Archean. *Earth and Planetary Science Letters*, 303(1), 121–132.

Crowe, S. A., Døssing, L. N., Beukes, N. J., et al. (2013). Atmospheric oxygenation three billion years ago. *Nature*, 501(7468), 535–538.

Dauphas, N., John, S. G., & Rouxel, O. (2017). Iron isotope systematics. *Reviews in Mineralogy and Geochemistry*, 82(1), 415–510.

Daye, M., Klepac-Ceraj, V., Pajusalu, M., et al. (2019). Light-driven anaerobic microbial oxidation of manganese. *Nature*, **576**(7786), 311–314.

Døssing, L. N., Dideriksen, K., Stipp, S. L. S., & Frei, R. (2011). Reduction of hexavalent chromium by ferrous iron: A process of chromium isotope fractionation and its relevance to natural environments. *Chemical Geology*, **285** (1), 157–166.

Farquhar, J., Zerkle, A. L., & Bekker, A. (2014). Geologic and geochemical constraints on Earth's early atmosphere. In H. D. Holland & K. K. Turekian, eds., *Treatise on Geochemistry* (2nd ed.), Oxford: Elsevier, pp. 91–138.

Fischer, W., & Knoll, A. H. (2009). An iron shuttle for deepwater silica in late Archean and early Paleoproterozoic iron formation. *Geological Society of America Bulletin*, **121**, 222–235.

Frei, R., Gaucher, C., Poulton, S. W., & Canfield, D. E. (2009). Fluctuations in Precambrian atmospheric oxygenation recorded by chromium isotopes. *Nature*, **461**(7261), 250–253.

Frost, C. D., von Blanckenburg, F., Schoenberg, R., et al. (2006). Preservation of Fe isotope heterogeneities during diagenesis and metamorphism of banded iron formation. *Contributions to Mineralogy and Petrology*, **153**(2), 211.

Galili, N., Shemesh, A., Yam, R., et al. (2019). The geologic history of seawater oxygen isotopes from marine iron oxides. *Science*, **365**(6452), 469–473.

Garvin, J., Buick, R., Anbar, A. D., et al. (2009). Isotopic evidence for an aerobic nitrogen cycle in the latest Archean. *Science*, **323**(5917), 1045–1048.

Goldberg, T., Archer, C., Vance, D., & Poulton, S. W. (2009). Mo isotope fractionation during adsorption to Fe (oxyhydr)oxides. *Geochimica et Cosmochimica Acta*, **73**(21), 6502–6516.

Goto, K. T., Sekine, Y., Shimoda, G., et al. (2020). A framework for understanding Mo isotope records of Archean and Paleoproterozoic Fe- and Mn-rich sedimentary rocks: Insights from modern marine hydrothermal Fe-Mn oxides. *Geochimica et Cosmochimica Acta*, **280**, 221–236.

Gross, G. A. (1980). A classification of iron formations based on depositional environments. *The Canadian Mineralogist*, **18**, 215–222.

Gueguen, B., Sorensen, J. V., Lalonde, S. V., et al. (2018). Variable Ni isotope fractionation between Fe-oxyhydroxides and implications for the use of Ni isotopes as geochemical tracers. *Chemical Geology*, **481**, 38–52.

Gumsley, A. P., Chamberlain, K. R., Bleeker, W., et al. (2017). Timing and tempo of the Great Oxidation Event. *Proceedings of the National Academy of Sciences of the United States of America*, **114**(8), 1811–1816.

Halevy, I., Alesker, M., Schuster, E. M., et al. (2017). A key role for green rust in the Precambrian oceans and the genesis of iron formations. *Nature Geoscience*, **10**(2), 135–139.

Hartman, H. (1984). The evolution of photosynthesis and microbial mats: A speculation on the banded iron formations. In Y. Cohen, R. Castenholz, & H. Halvorson, eds., *Microbial Mats: Stromatolites*, New York: Alan Liss, pp. 451–453.

Haugaard, R., Pecoits, E., Lalonde, S., et al. (2016). The Joffre banded iron formation, Hamersley Group, Western Australia: Assessing the palaeoenvironment through detailed petrology and chemostratigraphy. *Precambrian Research*, **273**, 12–37.

Hayashi, T., Tanimizu, M., & Tanaka, T. (2004). Origin of negative Ce anomalies in Barberton sedimentary rocks, deduced from La–Ce and Sm–Nd isotope systematics. *Precambrian Research*, **135**(4), 345–357.

Heard, A. W., Aarons, S. M., Hofmann, A., et al. (2021). Anoxic continental surface weathering recorded by the 2.95 Ga Denny Dalton Paleosol (Pongola Supergroup, South Africa). *Geochimica et Cosmochimica Acta*, **295**, 1–23.

Heard, A. W., Dauphas, N., Guilbaud, R., et al. (2020). Triple iron isotope constraints on the role of ocean iron sinks in early atmospheric oxygenation. *Science*, **370**, 446–449.

Heck, P. R., Huberty, J. M., Kita, N. T., et al. (2011). SIMS analyses of silicon and oxygen isotope ratios for quartz from Archean and Paleoproterozoic banded iron formations. *Geochimica et Cosmochimica Acta*, **75**(20), 5879–5891.

Heimann, A., Johnson, C. M., Beard, B. L., et al. (2010). Fe, C, and O isotope compositions of banded iron formation carbonates demonstrate a major role for dissimilatory iron reduction in ~2.5 Ga marine environments. *Earth and Planetary Science Letters*, **294**, 8–18.

Hoffman, P. F. (1987). Early Proterozoic foredeeps, foredeep magmatism, and Superior-type iron-formations of the Canadian Shield. In A. Kröner (Ed.) Geodynamics Series Volume 17: Proterozic Lithospheric Evolution, American Geophysical Union (AGU), Washington, DC, USA pp. 85–98.

Holland, H. D. (1973). The oceans: A possible source of iron in iron-formations. *Economic Geology*, **68**(7), 1169–1172.

James, H. L. (1954). Sedimentary facies of iron-formation. *Economic Geology*, **49**(3), 235–293.

Johnson, C. M., Beard, B. L., & Roden, E. E. (2008). The iron isotope fingerprints of redox and biogeochemical cycling in the modern and ancient Earth. *Annual Reviews in Earth and Planetary Sciences*, **36**, 457–493.

Johnson, C. M., Beard, B. L., & Weyer, S. (2020). *Iron Geochemistry: An Isotopic Perspective*, Cham: Springer International.

Johnson, J. E., Webb, S. M., Thomas, K., et al. (2013). Manganese-oxidizing photosynthesis before the rise of cyanobacteria. *Proceedings of the National Academy of Sciences*, 110(28), 11238–11243.

Jones, C., Nomosatryo, S., Crowe, S. A., et al. (2015). Iron oxides, divalent cations, silica, and the early earth phosphorus crisis. *Geology*, 43(2), 135–138.

Kappler, A., Pasquero, C., Konhauser, K. O., & Newman, D. K. (2005). Deposition of banded iron formations by anoxygenic phototrophic Fe(II)-oxidizing bacteria. *Geology*, 33(11), 865–868.

Kendall, B., Dahl, T. W., & Anbar, A. D. (2017). The stable isotope geochemistry of molybdenum. *Reviews in Mineralogy and Geochemistry*, 82(1), 683–732.

Knauth, L. P. (2005). Temperature and salinity history of the Precambrian ocean: Implications for the course of microbial evolution. In N. Noffke, ed., *Geobiology: Objectives, Concepts, Perspectives*, Amsterdam: Elsevier, pp. 53–69.

Konhauser, K. O., Hamade, T., Raiswell, R., et al. (2002). Could bacteria have formed the Precambrian banded iron formations? *Geology*, 30(12), 1079–1082.

Konhauser, K. O., Lalonde, S. V., Amskold, L., & Holland, H. D. (2007). Was there really an Archean phosphate crisis? *Science*, 315(5816), 1234.

Konhauser, K. O., Lalonde, S. V., Planavsky, N. J., et al. (2011). Aerobic bacterial pyrite oxidation and acid rock drainage during the Great Oxidation Event. *Nature*, 478(7369), 369–373.

Konhauser, K. O., Pecoits, E., Lalonde, S. V., et al. (2009). Oceanic nickel depletion and a methanogen famine before the Great Oxidation Event. *Nature*, 458(7239), 750–753.

Konhauser, K. O., Planavsky, N. J., Hardisty, D. S., et al. (2017). Iron formations: A global record of Neoarchaean to Palaeoproterozoic environmental history. *Earth-Science Reviews*, 172, 140–177.

Konhauser, K. O., Robbins, L. J., Alessi, D. S., et al. (2018). Phytoplankton contributions to the trace-element composition of Precambrian banded iron formations. *GSA Bulletin*, 130(5–6), 941–951.

Krapež, B., Barley, M. E., & Pickard, A. L. (2003). Hydrothermal and resedimented origins of the precursor sediments to banded iron formation: Sedimentological evidence from the early Palaeoproterozoic Brockman supersequence of Western Australia. *Sedimentology*, 50, 979–1011.

Lau, K. V., Romaniello, S. J., & Zhang, F. (2019). The uranium isotope paleoredox proxy. In T. Lyons, A. Turchyn & C. Reinhard (Eds.), *Elements in Geochemical Tracers in Earth System Science*, Cambridge University Press Cambridge, UK. pp. 1–28.

Li, W., Huberty, J. M., Beard, B. L., et al. (2013). Contrasting behavior of oxygen and iron isotopes in banded iron formations revealed by in situ isotopic analysis. *Earth and Planetary Science Letters*, **384**, 132–143.

Liljestrand, F. L., Knoll, A. H., Tosca, N. J., et al. (2020). The triple oxygen isotope composition of Precambrian chert. *Earth and Planetary Science Letters*, **537**, 116167.

Mänd, K., Lalonde, S. V., Robbins, L. J., et al. (2020). Palaeoproterozoic oxygenated oceans following the Lomagundi–Jatuli Event. *Nature Geoscience*, **13**, 302–306.

Mathur, R., Brantley, S., Anbar, A., et al. (2010). Variation of Mo isotopes from molybdenite in high-temperature hydrothermal ore deposits. *Mineralium Deposita*, **45**(1), 43–50.

Mloszewska, A. M., Pecoits, E., Cates, N. L., et al. (2012). The composition of Earth's oldest iron formations: The Nuvvuagittuq Supracrustal Belt (Québec, Canada). *Earth and Planetary Science Letters*, **317–318**, 331–342.

Morris, R. C. (1993). Genetic modelling for banded iron-formation of the Hamersley Group, Pilbara Craton, Western Australia. *Precambrian Research*, **60**(1), 243–286.

Moynier, F., Vance, D., Fujii, T., & Savage, P. (2017). The isotope geochemistry of zinc and copper. *Reviews in Mineralogy and Geochemistry*, **82**(1), 543–600.

Muhling, J. R., & Rasmussen, B. (2020). Widespread deposition of greenalite to form banded iron formations before the Great Oxidation Event. *Precambrian Research*, **339**, 105619.

Murray, K. J., Mozafarzadeh, M. L., & Tebo, B. M. (2005). Cr(III) oxidation and Cr toxicity in cultures of the manganese(II)-oxidizing Pseudomonas putida strain gb-1. *Geomicrobiology Journal*, **22**(3–4), 151–159.

Nakada, R., Takahashi, Y., & Tanimizu, M. (2016). Cerium stable isotope ratios in ferromanganese deposits and their potential as a paleo-redox proxy. *Geochimica et Cosmochimica Acta*, **181**, 89–100.

Ossa Ossa, F., Eickmann, B., Hofmann, A., et al. (2018). Two-step deoxygenation at the end of the Paleoproterozoic Lomagundi Event. *Earth and Planetary Science Letters*, **486**, 70–83.

Ostrander, C. M., Kendall, B., Olson, S. L., et al. (2020). An expanded shale δ^{98}Mo record permits recurrent shallow marine oxygenation during the Neoarchean. *Chemical Geology*, **532**, 119391.

Ostrander, C. M., Nielsen, S. G., Owens, J. D., et al. (2019). Fully oxygenated water columns over continental shelves before the Great Oxidation Event. *Nature Geoscience*, **12**(3), 186.

Partin, C. A., Lalonde, S. V., Planavsky, N. J., et al. (2013). Uranium in iron formations and the rise of atmospheric oxygen. *Chemical Geology*, **362**, 82–90.

Percak-Dennett, E. M., Beard, B. L., Xu, H., et al. (2011). Iron isotope fractionation during microbial dissimilatory iron oxide reduction in simulated Archaean seawater. *Geobiology*, **9**(3), 205–220.

Perry, E. C. (1967). The oxygen isotope chemistry of ancient cherts. *Earth and Planetary·Science Letters*, **3**, 62–66.

Petrash, D. A., Robbins, L. J., Shapiro, R. S., et al. (2016). Chemical and textural overprinting of ancient stromatolites: Timing, processes, and implications for their use as paleoenvironmental proxies. *Precambrian Research*, **278**, 145–160.

Pinti, D. L., Hashizume, K., & Matsuda, J. (2001). Nitrogen and argon signatures in 3.8 to 2.8 Ga metasediments: Clues on the chemical state of the Archean ocean and the deep biosphere. *Geochimica et Cosmochimica Acta*, **65**(14), 2301–2315.

Planavsky, N. J., Asael, D., Hofmann, A., et al. (2014). Evidence for oxygenic photosynthesis half a billion years before the Great Oxidation Event. *Nature Geoscience*, **7**(4), 283–286.

Planavsky, N. J., Bekker, A., Rouxel, O. J., et al. (2010a). Rare earth element and yttrium compositions of Archean and Paleoproterozoic Fe formations revisited: New perspectives on the significance and mechanisms of deposition. *Geochimica et Cosmochimica Acta*, **74**(22), 6387–6405.

Planavsky, N. J., Robbins, L. J., Kamber, B. S., & Schoenberg, R. (2020). Weathering, alteration and reconstructing Earth's oxygenation. *Interface Focus*, **10**, 20190140.

Planavsky, N. J., Rouxel, O. J., Bekker, A., et al. (2010b). The evolution of the marine phosphate reservoir. *Nature*, **467**(7319), 1088–1090.

Planavsky, N. J., Rouxel, O. J., Bekker, A., et al. (2012). Iron isotope composition of some Archean and Proterozoic iron formations. *Geochimica et Cosmochimica Acta*, **80**, 158–169.

Pons, M.-L., Fujii, T., Rosing, M., et al. (2013). A Zn isotope perspective on the rise of continents. *Geobiology*, **11**(3), 201–214.

Posth, N., Hegler, F., Konhauser, K. O., & Kappler, A. (2008). Alternating Si and Fe deposition caused by temperature fluctuations in Precambrian oceans. *Nature Geoscience*, **1**, 703–707.

Rasmussen, B., Muhling, J. R., Suvorova, A., & Krapež, B. (2017). Greenalite precipitation linked to the deposition of banded iron formations downslope from a late Archean carbonate platform. *Precambrian Research*, **290**, 49–62.

Raye, U., Pufahl, P. K., Kyser, T. K., et al. (2015). The role of sedimentology, oceanography, and alteration on the δ^{56}Fe value of the Sokoman iron formation, Labrador Trough, Canada. *Geochimica et Cosmochimica Acta*, **164**, 205–220.

Robbins, L. J., Funk, S. P., Flynn, S. L., et al. (2019a). Hydrogeological constraints on the formation of Palaeoproterozoic banded iron formations. *Nature Geoscience*, **12**(7), 558–563.

Robbins, L. J., Konhauser, K. O., Warchola, T. J., et al. (2019b). A comparison of bulk versus laser ablation trace element analyses in banded iron formations: Insights into the mechanisms leading to compositional variability. *Chemical Geology*, **506**, 197–224.

Robbins, L. J., Lalonde, S. V., Planavsky, N. J., et al. (2016). Trace elements at the intersection of marine biological and geochemical evolution. *Earth-Science Reviews*, **163**, 323–348.

Robbins, L. J., Lalonde, S. V., Saito, M. A., et al. (2013). Authigenic iron oxide proxies for marine zinc over geological time and implications for eukaryotic metallome evolution. *Geobiology*, **11**(4), 295–306.

Robbins, L. J., Swanner, E. D., Lalonde, S. V., et al. (2015). Limited Zn and Ni mobility during simulated iron formation diagenesis. *Chemical Geology*, **402**, 30–39.

Robert, F., & Chaussidon, M. (2006). A palaeotemperature curve for the Precambrian oceans based on silicon isotopes in cherts. *Nature*, **443**(7114), 969–972.

Rouxel, O. J., Bekker, A., & Edwards, K. J. (2005). Iron isotope constraints on the Archean and Paleoproterozoic ocean redox state. *Science*, **307**(5712), 1088–1091.

Rouxel, O. J., & Luais, B. (2017). Germanium isotope geochemistry. *Reviews in Mineralogy and Geochemistry*, **82**(1), 601–656.

Rudnick, R. L., & Gao, S. (2014). Composition of the continental crust. In H. D. Holland & K. K. Turekian (Eds.) *Treatise on Geochemistry*, Amsterdam: Elsevier, pp. 1–51.

Schad, M., Halama, M., Bishop, B., et al. (2019). Temperature fluctuations in the Archean ocean as trigger for varve-like deposition of iron and silica minerals in banded iron formations. *Geochimica et Cosmochimica Acta*, **265**, 386–412.

Scott, C., Lyons, T. W., Bekker, A., et al. (2008). Tracing the stepwise oxygenation of the Proterozoic ocean. *Nature*, **452**(7186), 456–459.

Skomurski, F. N., Ilton, E. S., Engelhard, M. H., et al. (2011). Heterogeneous reduction of U^{6+} by structural Fe^{2+} from theory and experiment. *Geochimica et Cosmochimica Acta*, **75**(22), 7277–7290.

Smith, A. J. B., Beukes, N. J., & Gutzmer, J. (2013). The composition and depositional environments of Mesoarchean iron formations of the West Rand Group of the Witwatersrand Supergroup, South Africa. *Economic Geology*, **108**, 111–134.

Smith, A. J. B., Beukes, N. J., Gutzmer, J., et al. (2017). Oncoidal granular iron formation in the Mesoarchaean Pongola Supergroup, southern Africa: Textural and geochemical evidence for biological activity during iron deposition. *Geobiology*, **15**, 731–749.

Steinhoefel, G., von Blanckenburg, F., Horn, I., et al. (2010). Deciphering formation processes of banded iron formations from the Transvaal and the Hamersley successions by combined Si and Fe isotope analysis using UV femtosecond laser ablation. *Geochimica et Cosmochimica Acta*, **74**(9), 2677–2696.

Stüeken, E. E., Kipp, M. A., Koehler, M. C., & Buick, R. (2016). The evolution of Earth's biogeochemical nitrogen cycle. *Earth-Science Reviews*, **160**, 220–239.

Swanner, E. D., Mloszewska, A. M., Cirpka, O. A., et al. (2015). Modulation of oxygen production in Archaean oceans by episodes of Fe(II) toxicity. *Nature Geoscience*, **8**(2), 126–130.

Swanner, E. D., Planavsky, N. J., Lalonde, S. V., et al. (2014). Cobalt and marine redox evolution. *Earth and Planetary Science Letters*, **390**, 253–263.

Thibon, F., Blichert-Toft, J., Albarede, F., et al. (2019). A critical evaluation of copper isotopes in Precambrian iron formations as a paleoceanographic proxy. *Geochimica et Cosmochimica Acta*, **264**, 130–140.

Thoby, M., Konhauser, K. O., Fralick, P. W., et al. (2019). Global importance of oxic molybdenum sinks prior to 2.6 Ga revealed by the Mo isotope composition of Precambrian carbonates. *Geology*, **47**(6), 559–562.

Thompson, K. J., Kenward, P. A., Bauer, K. W., et al. (2019). Photoferrotrophy, deposition of banded iron formations, and methane production in the Archean oceans. *Science Advances*, **5**, eaav2869.

Trendall, A. F., & Blockey, J. (1970). The iron formations of the Precambrian Hamersley Group, Western Australia with special reference to the associated crocidolite. *Western Australia Geological Survey Bulletin*, **119**, 1–366.

Trower, E. J., & Fischer, W. W. (2019). Precambrian Si isotope mass balance, weathering, and the significance of the authigenic clay silica sink. *Sedimentary Geology*, **384**, 1–11.

Urey, H. C. (1947). The thermodynamic properties of isotopic substances. *Journal of the Chemical Society (Resumed)*, **1**, 562–581.

Wang, C., Konhauser, K. O., & Zhang, L. (2015). Depositional environment of the Paleoproterozoic Yuanjiacun banded iron formation in Shanxi Province, China. *Economic Geology*, **110**, 1515–1539.

Wang, X., Planavsky, N. J., Hofmann, A., et al. (2018). A Mesoarchean shift in uranium isotope systematics. *Geochimica et Cosmochimica Acta*, **238**, 438–452.

Warke, M. R., Rocco, T. D., Zerkle, A. L., et al. (2020). The Great Oxidation Event preceded a Paleoproterozoic 'snowball Earth'. *Proceedings of the National Academy of Sciences*, **117**(24), 13314–13320.

Wasylenki, L. E., Rolfe, B. A., Weeks, C. L., et al. (2008). Experimental investigation of the effects of temperature and ionic strength on Mo isotope fractionation during adsorption to manganese oxides. *Geochimica et Cosmochimica Acta*, **72**(24), 5997–6005.

Wei, W., Klaebe, R., Ling, H.-F., et al. (2020). Biogeochemical cycle of chromium isotopes at the modern Earth's surface and its applications as a paleo-environment proxy. *Chemical Geology*, **541**, 119570.

Xu, W., Zhu, J.-M., Johnson, T. M., et al. (2020). Selenium isotope fractionation during adsorption by Fe, Mn and Al oxides. *Geochimica et Cosmochimica Acta*, **272**, 121–136.

Acknowledgements

Kaarel Mänd would like to acknowledge financial support from the Ministry of Education and Research of Estonia (mobility grant within the Archimedes Foundation's Kristjan Jaak Scholarship program), as well as UAlberta North and the Ashley and Janet Cameron Fund of the University of Alberta. Kurt O. Konhauser was supported by the Natural Sciences and Engineering Research Council of Canada Discovery grant (RGPIN-165831).

Cambridge Elements ☰

Geochemical Tracers in Earth System Science

Timothy Lyons

University of California

Timothy Lyons is a distinguished professor of Biogeochemistry in the Department of Earth Sciences at the University of California, Riverside. He is an expert in the use of geochemical tracers for applications in astrobiology, geobiology, and Earth history. Professor Lyons leads the 'Alternative Earths' team of the NASA Astrobiology Institute and the Alternative Earths Astrobiology Center at UC Riverside.

Alexandra Turchyn

University of Cambridge

Alexandra Turchyn is a university reader in Biogeochemistry in the Department of Earth Sciences at the University of Cambridge. Her primary research interests are in isotope geochemistry and the application of geochemistry to interrogate modern and past environments.

Chris Reinhard

Georgia Institute of Technology

Chris Reinhard is an assistant professor in the Department of Earth and Atmospheric Sciences at the Georgia Institute of Technology. His research focuses on biogeochemistry and paleoclimatology, and he is an Institutional PI on the 'Alternative Earths' team of the NASA Astrobiology Institute.

About the Series

This innovative series provides authoritative, concise overviews of the many novel isotope and elemental systems that can be used as 'proxies' or 'geochemical tracers' to reconstruct past environments over thousands to millions to billions of years – from the evolving chemistry of the atmosphere and oceans to their cause-and-effect relationships with life.

Covering a wide variety of geochemical tracers, the series reviews each method in terms of the geochemical underpinnings, the promises and pitfalls, and the 'state-of-the-art' and future prospects, providing a dynamic reference resource for graduate students, researchers and scientists in geochemistry, astrobiology, paleontology, paleoceanography and paleoclimatology.

The short, timely, broadly accessible papers provide much-needed primers for a wide audience – highlighting the cutting-edge of both new and established proxies as applied to diverse questions about Earth system evolution over wide-ranging time scales.

Cambridge Elements $\equiv$

Geochemical Tracers in Earth System Science

Printed in the United States
by Baker & Taylor Publisher Services

Printed in the United States
by Baker & Taylor Publisher Services